"Life Finds a Way"

A Paradigm Shift of Emerging, Disruptive Technologies for Sheltering

Sheltering Vehicle Minimum Viable Product (MVP)

About this Book

The purpose of this book is to generate discussion of a Sheltering Vehicle Minimal Viable Product (MVP [bib#1]) with the intended outcome of this book being the prototyping of a Sheltering Vehicle MVP and the supporting infrastructure.

As such feel free to crowdsource and distribute this book to anyone you think might be interested in the general topics discussed in this book. Feel free to contribute to this book by adding information and adapting this book as you see fit. As you do this, it is probably worthwhile to keep a list of those who have contributed to the book and those interested in the topic in case anyone in the process wants to follow-up with them in any way. A list is started below:

Current Contributors to this Book and Working Group are:
Margaret Kilduff, Ph.D, Professor Emerita, Rutgers University and

Associated Sheltering Vehicle Working Group Members. Contact
Information: meg.kilduff@rutgers.edu or margaret.kilduff@gmail.com

ISBN-13: 978-1-7336928-6-1
Imprint: Margaret Kilduff, Independent Publishing
Version 1.0, June 2020

Margaret Kilduff, Ph.D.
Professor Emerita
Rutgers University

Note: *The photograph used for the cover of the book was produced by Elgardo Lagmay and downloaded from Unsplash (https://unsplash.com/@fotograpi)*

Table of Contents

Note: The above picture is adapted from the one by Nathan Dumlao *found on Unsplash (https://unsplash.com/photos/XZToRBmGPcs)*

Introduction

The phrase "life finds a way" is one of the iconic lines from Jurassic Park. In the book, it is part of the following thought [bib#3] by mathematician Ian Malcolm:

> "*Broadly speaking, the ability of the park is to control the spread of life forms. Because the history of evolution is that life escapes all barriers. Life breaks free. Life expands to new territories. Painfully, perhaps even dangerously. But life finds a way.*"

And one way that life has found for at least one million Americans is the abandonment of permanent, traditional (fixed) sheltering (housing) in favor of residence (sheltering) in a vehicle. This vehicle resident *"lifestyle technology"* has broken free and can be considered emerging and disruptive. It is a paradigm shift. It is a *"new normal"* option for many people.

A classic text on paradigms and paradigm shifts is *The Structure of Scientific Revolutions* [bib#4] by Thomas Kuhn. A classic quote on the importance of having the correct paradigm is one by Ludwig Wittgenstein:

> *"A man will be imprisoned in a room* [bib#5] *with a door that's unlocked and opens inwards; as long as it does not occur to him to pull rather than push."*

Wittgenstein was a fascinating philosopher. An excellent source for more information about him is a YouTube lecture entitled *"Wittgenstein His Life and Philosophy"* [bib#6].

Campground [bib#7]

Purpose and Intended Outcome of this Book

The purpose of this book is to generate discussion of a Sheltering Vehicle Minimal Viable Product (MVP [bib#1]) which supports the *"new normal"* paradigm shift in housing/sheltering options. The Sheltering Vehicle supports the emerging, disruptive Vehicle Resident lifestyle and is envisioned to be, in and of itself, an emerging, disruptive technology. The intended outcome of this book is the prototyping of a Sheltering Vehicle MVP and the supporting infrastructure.

One thought is that communities should provide Sheltering Vehicles to a subset of the homeless population. The Sheltering Vehicle concept is better than sheds, tents, and most high density shelters. It is a better concept for many reasons which are discussed later in the book.

Oregon Trail'R Teardrop Trailer [bib#8]

One reason worth noting now, however, is that many people consider a vehicle such as the Sheltering Vehicle to provide more safety, privacy, autonomy as well as a greater sense of independence, competency, and self-sufficiency than do sheds, tents, and most high density shelters.

Another reason is that it is easier to clean around and under a Sheltering Vehicle than other options (e.g., tent) which eliminates some of the health and sanitary issues (e.g., rats, typhus [bib#9]) found with some homeless encampments.

The Sheltering Vehicles are also dynamic – they can be easily moved – in ways that tents, sheds, etc. cannot. This not only facilitates cleaning the general area, but also adjustments as local situations change.

Almost everyone should find this book useful and interesting. However, those with an interest in emerging, disruptive technologies for vehicles and/or housing/sheltering may find the book more useful and interesting. Those interested in new solutions for the problems of homelessness in the United States should also find this book interesting.

The sections of this book are:

Emerging, Disruptive Vehicle Technology in a Paradigm Shift

Vehicles as a component of emerging, disruptive technologies are not new. There is great interest and general public awareness of the paradigm shift in new forms of transportation and new ways to fuel that transportation (e.g., fast trains, electric cars, cars which run on natural gas).

A large, less noticed change is the relationship of vehicles to a major demographic paradigm shift happening in the United States with respect to how people are choosing to live. Many people are abandoning permanent, traditional housing; apartments and houses.

Many find such housing out-of-reach. Situations involving employment, education, and economic disruptions do not allow for a commitment to permanent, traditional housing. A commitment to home ownership – a mortgage – and multi-year rental agreements requires not only financial resources, but an expectation that you will be in a particular area for a fairly long period of time.

This may have been a reasonable expectation in the past, but today's employees generally do not have such an expectation. They are often not sure if their job will exist long term and/or whether they wish to work for one company or in one geographic area long term. Some prefer a life free of permanent, traditional housing.

Many of these people become Vehicle Residents. As stated earlier, this Vehicle Resident *"lifestyle technology"* is emerging and disruptive. It is a paradigm shift. How we as a society respond to this shift will impact the quality of life not only for Vehicle Residents, but also those communities based in permanent, traditional housing. As a society we can either ignore this shift – at our peril – or acknowledge the shift and engineer solutions which benefit the quality of life for all.

Ignoring the shift imprisons us in a room. It results in tensions between the Vehicle Residents and those who live in fixed (non-vehicle) resident communities. The tensions usually result from a conflict over resources. Often the source of the conflict is that Vehicle Residents use the community resources of the fixed resident community without paying for their use – and without having paid for either their development or for their maintenance.

The cost of developing and maintaining the fixed resident resources were and are paid by the fixed resident community. Vehicle Residents use resources which the fixed resident community does not want them to use. Acknowledging the shift means working to ensure that the Vehicle Resident lifestyle interfaces well with the fixed (non-vehicle) resident communities.

There is a need for recognition and acceptance of this demographic shift and a combined response to it. There is a need to establish a functioning Vehicle Resident community which interfaces well with functioning and recognized fixed resident communities. There are two aspects to this lifestyle technology interface:

1. A Sheltering Vehicle which meets the needs of both the Vehicle Resident and the communities with which the Sheltering Vehicle comes in contact; and

2. Vehicle Resident community resources (support facilities, amenities) external to the Sheltering Vehicle which are accessible to and meet the needs of the Vehicle Resident and which do not conflict with the other communities with which the Vehicle Resident comes in contact.

Elio Motors $7,450 Base Price [bib#10]

Sheltering Vehicle Minimum Viable Product (MVP)

Each aspect of this lifestyle technology interface could benefit from the acknowledgement of a need for and the development of a Minimal Viable Product (MVP [bib#1]). Such a MVP in each aspect (Sheltering Vehicle, Vehicle Resident community resources) would advance societal acknowledgement of the lifestyle technology paradigm shift and reduce tensions resulting from it.

This book is intended to serve as background for the development of a MVP in each aspect. However, the book is more focused on the MVP Sheltering Vehicle. Clearly, many people are currently living in vehicles (e.g., RVs, SUVs), however, the lifestyle technology paradigm shift most likely requires the development of an additional vehicle – a Sheltering Vehicle – one that contains the amenities of a shelter, not necessarily the amenities of permanent, traditional housing. Some suggested characteristics of such a Sheltering Vehicle are:

1. Relatively inexpensive to purchase (e.g., $7500 to buy) or rent/lease (can be purchased/leased/rented by individuals or by organizations on behalf of individuals such as organizations and government agencies which address homelessness)

2. Local traffic/road capable mainly (no need for the vehicle to be able to achieve highway speeds for extended periods) and street legal

3. Gas, electric, or hybrid powered, but inexpensive to operate (e.g., high MPG)

4. Easy and inexpensive to maintain (e.g. like the old Volkswagen Beetle, almost any owner could do basic maintenance)

5. Easy to drive (can be an autonomous vehicle)

6. Structured so that at least one person who is at least 6 feet tall can sleep comfortably

7. Some storage capability (e.g., for clothes)

8. Safe and secure for sleeping and stored items

9. No need for incorporated non-shelter amenities (e.g., no cooking facilities, no bathroom)

10. A new option for current and potential Vehicle Residents (is not intended to replace current options)

11. Aesthetically appealing (is not an eyesore in any community in which it is parked)

12. Unique, easily identifiable design characteristics to build community (e.g., similar to the iconic Airstream Community [bib#11]) and deter theft (e.g., similar to the design deterrent of bike-sharing bikes [bib#12])

13. Appealing to both those seeking to be a Vehicle Resident as well as those who are looking for a versatile, functional vehicle with no intention of being a Vehicle Resident (potential market both Vehicle Residents and those living in permanent, traditional housing)

As will be discussed later in this book, a possible new parking option for such a Sheltering Vehicle might be the parking lot of a currently abandoned/vacant mall. Community resources (support facilities, amenities such as bathrooms, showers, washing machines) for those Vehicle Residents in the Sheltering Vehicles could be located in the abandoned mall building or in the parking lot.

Vehicle Residents: Nomads/Travelers and Geographically Local

The development of a MVP Sheltering Vehicle requires an understanding of the Vehicle Resident market. It requires an understanding of who Vehicle Residents are. Vehicle Residents fall into two broad categories: 1) Nomads/Travelers and 2) Geographically Local. Although it should be noted that the Sheltering Vehicle may have widespread appeal and be of interest to those in permanent, traditional (fixed) housing looking for a versatile, functional vehicle.

Nomads/Travelers are the people who have chosen to lead a nomadic, traveling lifestyle (e.g., full-time RVers). They live in vehicles to facilitate easy, and usually widespread, travel. It is estimated that one million people [bib#13] live and travel in a RV or campervan full-time or close to full-time. They usually park in areas intended for the overnight use of those living in vehicles (e.g., RV campgrounds) and usually do not stay in one local geographic area for any extended period of time; "*home*" is on-the-road in their vehicle, not in one local geographic area. The travel locations of some people in this group are determined by employment; their work locations regularly change and they travel to the locations where they work. Other people determine their travel locations by their own personal interests.

The Geographically Local Vehicle Residents tend to stay in one rather local geographic area. They do not choose to live in a vehicle to facilitate travel. For the Geographically Local Vehicle Resident, living in a vehicle is a way to stay in one local geographic area when other sheltering/housing options (e.g., permanent, fixed housing) are either not available or not feasible. Often these Vehicle Residents are called *"homeless"* by society, but this is a misnomer. Generally speaking, Geographically Local Vehicle Residents do not consider themselves to be homeless. They have a *"home"* – their vehicle. In this book, a vehicle is defined as a structure on wheels with or without an engine (e.g., with an engine = car; without an engine = trailer).

It also seems that it is the wrong paradigm to consider them to be homeless. Geographically Local Vehicle Residents have more in common with the

Nomads/Travelers than they do with those who are homeless. Although the paradigm which considers Geographically Local Vehicle Residents to be homeless is incorrect in light of the *"new normal*", this book discusses the homeless and homelessness because many still use this paradigm.

Overview of Homelessness in the United States

The United States Department of Housing and Urban Development (HUD [bib#14]) produces the Annual Homeless Assessment Report (AHAR [bib#15]) for Congress. The 2018 report [bib#16] states that:

"***Homeless*** *describes a person who lacks a fixed, regular, and adequate nighttime residence.*"

"***Unsheltered Homelessness*** *refers to people whose primary nighttime location is a public or private place not designated for, or ordinarily used as, a regular sleeping accommodation for people (for example, the streets, vehicles, or parks).*"

HUD estimates that approximately 550,000 people were homeless on any given night [bib#17] in 2018. Of these, approximately 200,000 are unsheltered while approximately 350,000 are sheltered (e.g., in emergency shelters or transitional housing).

HUD states that approximately 60 percent of all homeless people are male (men or boys) while 70 percent of the unsheltered homeless are male. Approximately 9 percent of all homeless people are veterans. Individuals comprise 67 percent of the entire homeless population while 33 percent are part of a homeless family.

Half of all homeless people are found in one of five states: California, New York, Florida, Texas, or Washington. California also has more than half of all the unsheltered homeless in the United States. Those states where more than half of all homeless are unsheltered are: California, Oregon, Nevada, and Hawaii.

Federal spending on homelessness comes through many federal agencies. The total annual federal expenditure across all agencies is approximately $4.75 billion [bib#18]. The two federal agencies with the largest annual expenditure on homeless-related programs are:

1. HUD which appropriates approximately $2.6 billion for Homeless Assistance Grants [bib#19]; and

2. Veterans Affairs which appropriates approximately $1.8 billion [bib#20] for programs directed at homeless veterans [bib#21] and those veterans at-risk [bib#22] of becoming homeless.

The homelessness efforts of the different federal agencies are coordinated by the United States Interagency Council on Homelessness *[bib#23]* which works *"in close partnership with senior leaders across our 19 federal member agencies*". Additional expenditures on the homeless and homelessness are appropriated by state and local government sources as well as non-profit organizations (e.g., religious institutions, Salvation Army).

Additional resources for information on the homeless and homelessness in the United States are:

- Homelessness in the United States, Wikipedia [bib#24]
- National Alliance to End Homelessness [bib#25]
- Coalition for the Homeless [bib#26]
- Salvation Army, Provide Shelter [bib#27]
- List of homelessness organizations, Wikipedia [bib#28]
- Is 'Homeless' The Right Word For Those Living On The Street? [bib#29]
- The Homeless Industrial Complex [bib#30]
- Disguising a Public Safety Crisis as a Housing Crisis [bib#31]

Unsheltered Homeless

"***Unsheltered Homelessness*** *refers to people whose primary nighttime location is a public or private place not designated for, or ordinarily used as, a regular sleeping accommodation for people (for example, the streets, vehicles, or parks).*"

Many of those currently defined as unsheltered homeless live in vehicles (cars, vans, RVs). Many choose this option [bib#32] because "*With doors that lock and a watertight roof, cars offer a sense of safety, privacy, and autonomy—even if parked on a public street—that tents and emergency shelters do not.* " Clearly, to choose the vehicle option, a homeless person needs to have the financial resources or assistance to acquire a vehicle.

It is reasonable to assume that many of those considered to be unsheltered homeless living on the street (e.g., in a tent, in a makeshift shelter, on public transportation) would choose a vehicle for shelter over their current living situation if the opportunity existed. It is also reasonable to assume that some of the sheltered homeless (e.g., in emergency shelters, in temporary housing) might also choose the vehicle option if personal resources allowed it – for the same reasons of safety, privacy, autonomy as well as a greater sense of independence, competency, and self-sufficiency.

Safety, privacy, autonomy, a sense of independence, competency, and self-sufficiency are absolutely critical for any living situation. Many homeless do not choose to go to a homeless shelter because they feel the shelters lack these attributes – shelters are not safe, do not allow for a sense of autonomy, etc. A vehicle satisfies all of these issues.

Those considered the unsheltered homeless in vehicles in the old paradigm and the Geographically Local Vehicle Residents in the new paradigm can be organized into two broad groups:

1. Those who are connected to a wider stable community through employment or school (i.e., students) and lead fairly structured and organized lives; and

2. Those who are unable to connect to a wider stable community because of personal characteristics which make leading a structured and organized life difficult (e.g., mental illness, addiction).

Each of these groups tends to stay in one local geographic area; "*home*" is in a local geographic area. For these unsheltered homeless, the vehicle is usually a Sheltering Vehicle.

Clearly, those in the second group need a wider range of community resources (support facilities, amenities) external to the Sheltering Vehicle than do those in the first group.

An example of those is the first group is college students. In the California State University (CSU) system, approximately 11 percent of students are homeless [bib#33], many of whom live in their vehicles.. San Jose State University (SJSU) has the highest number of homeless students [bib#34] in the CSU system (approximately 13 percent of the student body; more than 4,000 homeless students).

Before moving on to the next section it is worthwhile to briefly note some of the small fixed (non-vehicle) housing which has been used in the past and some that are in use today. Some of these options may provide inspiration for the Sheltering Vehicle MVP.

One small fixed housing option found commonly in the past, but not as common today is the Single Room Occupancy (SRO) [bib#35]. The SRO "*is a form of housing that is typically aimed at residents with low or minimal incomes who rent small, furnished single rooms with a bed, chair, and sometimes a small desk ... tenants share a kitchen, toilets or bathrooms*". Some people consider coliving units such as Starcity [bib#36] , Common [bib#37], and WeLive [bib#38] to be just a new, modern form of the SRO.

Another small fixed housing option is the capsule or pod hotel [bib#39] which *"is a type of hotel developed in Japan that features a large number of small bed-sized rooms known as capsules."* One somewhat upscale example is First Cabin Tsukiji [bib#40]. Another interesting Japanese example is the internet café [bib#41] which often functions as a part-time *"home"* for many. So many people were and are doing this that one internet café operator (Manboo) formalized it with the introduction of the Net Room [bib#42] – an internet café with private rooms and amenities such as washing machines.

It should also be noted that some cities are experimenting with villages of fixed backyard sheds [bib#43] and pop-up shelters [bib#44] to house homeless people

Additional resources for information on unsheltered homelessness and related topics are:

- Homes on Wheels: Meet the People Who Live in RVs on San Francisco Streets [bib#45]
- More Americans are forced to "reside" in their vehicles [bib#46]
- America's Housing Crisis Is Forcing More People To Live In Vehicles [bib#47]
- 'It's the New Form of Affordable Housing': More People Are Living in Their Cars [bib#48]
- For Countless Americans, Living In Their Vehicles Has Become "The New Normal" [bib#49]
- Mountain View Vehicle Residents throw party for public [bib#50]
- San Jose State University has highest number of homeless students across CSU system [bib#51]
- Homeless Shelter [bib#52]
- Tent City, Wikipedia [bib#53]
- Is It a Good Idea to Give Tents to Homeless People? [bib#54]
- Tents-4-Homeless [bib#55]
- Weedpatch Camp [bib#56]

Summary of the Paradigm Shift: Comparison of the Old Paradigm with the New Paradigm

As was stated at the beginning of this book, there is a large demographic paradigm shift occurring in housing/sheltering. Americans are abandoning permanent, traditional (fixed) sheltering (housing) in favor of residence (sheltering) in a vehicle. It is a *"new normal"* option for many people. A summary of this paradigm shift can be found in the following table:

Unsheltered Homeless

Old Housing/ Sheltering Paradigm

- People living in the open (e.g., on the street) without shelter or in a tent or makeshift shelter

New Housing/Sheltering Paradigm

- Those whose personal situation is chaotic (e.g., suffer from mental illness) remain living in the open *(Unsheltered Homeless).*

- Those who are able to handle the living characteristics of a Sheltering Vehicle (e.g., self-sufficiency, autonomy) move to a Sheltering Vehicle *(Vehicle Resident).*

Sheltered Homeless

Old Housing/ Sheltering Paradigm

- People living in a homeless shelter or transitional housing (includes those in FEMA emergency housing)

New Housing/Sheltering Paradigm

- Those for whom a homeless shelter or transitional housing is the preferred option remain in that situation *(Sheltered Homeless).*

- Those for whom a homeless shelter or transitional housing is not a preferred option and who are able to handle the living characteristics of a Sheltering Vehicle (e.g., self-sufficiency, autonomy) move to a Sheltering Vehicle *(Vehicle Resident).*

Housed

Old Housing/ Sheltering Paradigm

- People living in permanent, traditional fixed housing (e.g., apartments, condominiums, houses)

New Housing/Sheltering Paradigm

- Those who prefer permanent, traditional fixed housing and are able to meet the requirements remain in the housing *(of course)*.

- Those who prefer permanent, traditional fixed housing, but are unable to meet the requirements become Vehicle Residents – either long or short term.

- Those who have grown weary of permanent, traditional fixed housing choose to become Vehicle Residents.

Parking for Those Living or Sheltering in Vehicles

Regardless of whether the person living in a vehicle is unsheltered homeless (Geographically Local Vehicle Resident) or a Nomad/Traveler, the person needs a place to park the vehicle. Parking is just one – although clearly an essential one – of the community resources (support facilities, amenities) needed by the Vehicle Resident. This parking resource needs to be accessible to and meet the needs of the Vehicle Resident and not conflict with the other communities with which the Vehicle Resident comes in contact.

The Nomads/Travelers usually have such a resource because they have the money, the information network, and the mobility to find and park in places which are accepting or even welcoming of those living in vehicles (e.g., RV campgrounds, public campgrounds). Often these parking spaces have a fee and also provide services (e.g., water, laundry facilities).

Although the cost per night in such parking spaces varies widely by location and type, it is estimated [bib#57] that the average cost of parking in a private RV campground is $38.50 per night (approximately $14,000 per year) while the average cost in a public campground (e.g., state park campground, national park campground) is $22.15 per night (approximately $8,000 per year).

Go RVing [bib#58] provides information on RV Parks and Campgrounds [bib#59] as well as where RVs are welcome overnight in public land campgrounds [bib#60]. There are over 13,000 privately owned RV parks/campgrounds and over 1,600 of the total 7,500 state parks in the United States cater to RVers [bib#61].

The Nomads/Travelers usually can afford to pay for parking overnight and factor that cost into their annual living expenses. Although, clearly, the Nomad/Traveler can and often will take advantage of any free, legal overnight parking to reduce "*housing costs*". Such places include, but are not limited to some Walmart, Cracker Barrel, Bass Pro Shop, Home Depot, and casino parking lots.

The unsheltered homeless (Geographically Local Vehicle Resident) living in vehicles, however, need free parking since they generally cannot afford the expense of paid parking. As a result, the unsheltered homeless living in vehicles tend to park wherever they can. Sometimes they can take advantage of free, legal overnight parking also used by Nomad/Traveler (e.g., in Walmart parking lots), however, they often park on the street. This often annoys local residents who view them as a drain on – and not contributing participants in – the local community especially if the unsheltered homeless person has a vehicle in very poor condition, has poor hygiene, or does not respect the private property of others in the neighborhood. As a result, many towns have criminalized vehicle residency.

Some towns and organizations have sought to solve the problem of tension between unsheltered homeless in vehicles and local housed residents by developing parking lots designated for the use of unsheltered homeless in vehicles. This effort is often called a Safe Parking program. The first official Safe Parking program is considered to be the one started in 2004 in Santa Barbara, CA by the New Beginnings Counseling Center [bib#62] . Two additional examples are the Safe Parking Program operated by Dreams for Change [bib#63] in San Diego, CA and Safe Parking LA [bib#64] in Los Angeles, CA.

Some colleges and universities are considering setting aside campus parking lots for overnight safe parking for unsheltered homeless students living in vehicles. The California Legislature Assembly Bill AB-302 Parking: homeless students [bib#65], for example, would authorize community colleges to set aside overnight parking facilities for unsheltered homeless students in vehicles (Geographically Local Vehicle Resident).

Information which might be relevant to the long term development of safe parking programs is the number of parking spaces in the United States and possible available parking lots. It is estimated that the United States has 2 billion parking spaces [bib#66], but only 250 million cars (a ratio of 8 parking spaces available for every car). Some of these parking spaces are in mall parking lots. And many of the malls are no longer in operation which means these parking spaces are not in current use at all. A mall no longer in operation are sometimes called a "*dead mall*" [bib#67]. It is estimated that there are approximately 450 dead malls [bib#68] in the United States. Many more are only partially occupied.

Additional resources for information on parking for unsheltered homeless in vehicles and related topics are:

- California bill would let homeless college students sleep in cars on campus amid housing crisis [bib#69]
- 50 Unconventional Free Overnight Parking Recommendations [bib#70]
- 8 Places to Park and Sleep in Your Car Overnight for Free (Legally) [bib#71]
- All About Overnight Parking [bib#72]
- Vanlifer's Guide to Sleeping, Camping, and Overnight Parking [bib#73]
- Finding Home in a Parking Lot [bib#74]
- Living Behind the Wheel [bib#75]
- Where to Sleep in Your Car When Homeless [bib#76]
- Parking Has Eaten American Cities [bib#77]
- Quantified Parking: Comprehensive Parking Inventories for Five U.S. Cities [bib#78]
- List of shopping malls in the United States [bib#79]
- Hidden in Plain Sight: Finding Safe Parking for Vehicle Residents [bib#80]
- Championing Seattle's Invisible Homeless [bib#81]
- Seattle's forgotten street community: UW anthropologist talks about the unique circumstances of vehicle residency [bib#82]
- Seattle Vehicular Residency Research Project 2012 Advisory Report [bib#83]

Resources Needed for Those Living or Sheltering in Vehicles

Although the parking resource is essential for those unsheltered homeless (Geographically Local Vehicle Resident) and Nomads/Travelers living in a vehicle, many additional community resources (support facilities, amenities) are also needed. As with the parking resource, these additional resources must be accessible to and meet the needs of the Vehicle Resident and not conflict with the other communities with which the Vehicle Resident comes in contact. These additional (non-parking) community resources include, but are not limited to:

- Physical/Permanent Address and Internet
- Personal and Community Hygiene
- Meals and Storage

Physical/Permanent Address and Internet

Many functions of daily life (e.g., RealID, voter registration, vehicle registration, insurance) require an established physical, permanent address. Establishing such a physical/permanent address is not an easy thing to do for either Nomads/Travelers or the unsheltered homeless. Some are fortunate enough to have a friend or relative who allows them to use that friend's/relative's physical/permanent address as their own. For others, the task of establishing such an address presents a very real challenge.

For Nomads/Travelers, however, there are organizations which help them establish a physical/permanent address. One such organization is EscapeesRVClub [bib#84] which provides information and resources to help people living on the road full-time establish a domicile [bib#85] (permanent address) in a state. Some states (e.g., Texas, Florida, South Dakota) are friendly toward Nomads/Travelers; the requirements for establishing residency (e.g., spending a certain amount of time in the state) are ones which Nomads/Travelers can reasonably meet. EscapeesRVClub can also provide a mailing address and mail forwarding service [bib#86]. There is also information on internet options [bib#87] for Nomads/Travelers.

Although Nomads/Travelers can usually secure access to a physical/permanent address, and internet (as well as a cell phone), such access is much more difficult or non-existent for most unsheltered homeless living in vehicles. The Geographically Local Vehicle Resident usually does not have the mobility to establish residency in one of the states accommodating to Vehicle Residents. They need a physical/permanent address in the state in which they are living in their vehicle. Some studies have found that many unsheltered homeless have a cell phone [bib#88] (e.g., a disposable cell phone), but lack internet access.

Additional resources for information are:

- HomelessnessAdvice.com: Advice When You Need It, Guides [bib#89]
- Vanlife How To: Your Complete Guide to Life on the Road [bib#90]
- Workamper News [bib#91]
- How to Establish State Residency if you Live in Your RV Full Time [bib#92]
- Domicile for Nomads: Establishing a Legal Residence on the Road [bib#93]
- Mail and Packages on the Road: A Guide for Vanlife Nomads [bib#94]
- Helping you Live a Mobile & Connected Lifestyle! [bib#95]
- RV Internet Access [bib#96]

Personal and Community Hygiene

Access to resources (support facilities, amenities) to maintain personal and community hygiene are critical. These resources include, but are not limited to: adequate bathroom facilities, showering facilities, laundry facilities, and trash facilities.

These are critical not only for the health and safety of the Vehicle Residents, but also the communities with which they come in contact. Conditions and diseases which develop and spread in communities without personal and community hygiene (e.g., rats, typhus [bib#9]) can devastate that community and the communities with which they come in contact.

Many Nomads/Travelers have adequate access to some of these facilities in their RVs/campervans and a support infrastructure to keep these facilities functioning (e.g., an infrastructure to obtain clean water, offload grey/dirty water, and throw out trash). Many campgrounds provide these facilities to those paying to stay at the campground. There are also guides for Nomads/Travelers [bib#97] on where to find these facilities while on the road.

One suggested showering option for Nomads/Travelers is obtaining a gym membership [bib#98] at a national chain such as Planet Fitness [bib#99] and 24 Hour Fitness [bib#100] and showering in the gym facilities. Costco even sells a discounted membership [bib#101] to 24 Hour Fitness

For unsheltered homeless living in vehicles (Geographically Local Vehicle Resident), however, access to such facilities can be very difficult. In response to this situation, some organizations bring mobile bathrooms, showering facilities, and laundry facilities to areas with many unsheltered homeless. Some of these organizations are:

- Marin Mobile Care: The Showers [bib#102]
- Lava Mae Mobile Hygiene [bib#103]
- Dignity on Wheels (Showers and Laundry) [bib#104]
- The Laundry Truck LA [bib#105]
- Homeward Alliance Mobile Laundry [bib#106]
- Revive and Refresh [bib#107]

It should be noted that it is easier to clean around and under a Sheltering Vehicle than other options (e.g., tent) provided to *"unsheltered homeless"* which eliminates some of the health and sanitary issues (e.g., rats, typhus [bib#9]) found with some homeless encampments. The Sheltering Vehicles are also dynamic – they can be easily moved – in ways that tents, sheds, etc. cannot. This not only facilitates cleaning the general area, but also adjustments as local situations change.

Additional resources for information are:

- Homeless Encampment Cleaning Company [bib#108]
- HomelessnessAdvice.com: Advice When You Need It, Guides [bib#89]
- Vanlife How To: Your Complete Guide to Life on the Road [bib#90]
- Portable Restroom Trailers Sales and Rentals [bib#109]
- Mobile Bathroom [bib#110]
- Luxury Restroom Trailers [bib#111]
- PortaPotty Rental Guide [bib#112]
- Dumpsters.com [bib#113]

Meals and Storage

Access to adequate meals and adequate storage for food and other items are essential for daily life. Many Nomads/Travelers can maintain adequate access to meals, food, and storage in their RVs/campervans. Many campgrounds provide cooking facilities to those paying to stay at the campground. Many Nomads/Travelers also use self-storage facilities. The average monthly cost for a self-storage unit [bib#114] is approximately $90. For unsheltered homeless living in vehicles (Geographically Local Vehicle Resident), however, access to meals, food, and storage can be very difficult.

Two organizational responses to the meal/food situation are the soup kitchen [bib#115] and food bank [bib#116]. A homeless response to storage is often the shopping cart. One often sees unsheltered homeless with shopping carts used for the storage of food and other items. However, secure, and individually dedicated local and easily accessible storage space in the form of bins or lockers would be better.

Additional resources for information are:

- HomelessnessAdvice.com: Advice When You Need It, Guides [bib#89]
- Vanlife How To: Your Complete Guide to Life on the Road [bib#90]
- Temporary Kitchens 123 [bib#117]
- U.S. Mobile Kitchens [bib#118]
- 10 Best Cargo and Carriers for Cars in 2019 [bib#119]

- Thule Rooftop Cargo Carriers [bib#120]
- Home Depot Storage Bins [bib#121]

Possible Inspirations/Foundations for the Sheltering Vehicle Minimum Viable Product

What does the Sheltering Vehicle Minimum Viable Product look like? That is up for discussion. What are your ideas? If you are interested in considering this issue, the following information may serve as inspiration or a foundation for your thinking.

There are many vehicles in which people currently live/shelter. A brief overview of some of these options might prove helpful in considering a Sheltering Vehicle Minimum Viable Product.

Clearly, the RV is designed for short-term and long term living for the Nomad/Traveler and is used by some Geographically Local Vehicle Residents. There are many RV options [bib#122] from which to choose. Some Nomads/Travelers and Geographically Local Vehicle Residents live in campervan [bib#123] vehicles. One example of a campervan is the Volkswagen Autosleeper Clubman GL [bib#124], which is a typical European campervan, shown below.

Volkswagen Autosleeper Clubman GL [bib#124]

RVs and campervans have engines. Some small, portable, living space vehicle options without engines are tiny houses [bib#125] and the classic teardrop trailer [bib#126] such as those sold by England's Teardrop Trailers [bib#127] and nuCamp [bib#128].. An example of a teardrop trailer [bib#129] is shown below.

An example of a teardrop trailer [bib#129]

There are even guides and aftermarket vendors for turning regular vans (e.g., Ford Transit Connect Cargo Van [bib#130], Ford Transit Connect Passenger Wagon [bib#131]) into campervans or cars into campervans. Examples include:

- Van Life Guide: How To Build A DIY Camper Van Conversion [bib#132]
- The Vanual Build Manual [bib#133]
- Build Your Van [bib#134]
- DIY camper van: 5 affordable conversion kits you can buy now [bib#135]
- Campervan Conversion Kit: 7 simple ways to DIY your van [bib#136]
- Ursa Minor Vehicles [bib#137] converts the Honda Element , Jeep Wrangler Unlimited, and Ford Transit Connect Into small campers

For the Sheltering Vehicle Minimum Viable Product, information on some innovative vehicle designs and efforts might also provide some inspiration and a foundation. Examples include research vehicles from Volkswagen [bib#138]; one such vehicle is the Volkswagen NILS [bib#139] shown below:

Volkswagen NILS [bib#139]

Another example is the vehicle from Elio Motors [bib#140]. A picture of the Elio Motors three-wheel vehicle [bib#141] ($7,450 base price) is shown below:

Elio Motors three-wheel vehicle [bib#141]

Two additional companies which might be of interest as inspiration and foundation for the Sheltering Vehicle Minimal Viable Product are Local Motors [bib#142] and Open Motors [bib#143]. Open Motors (formerly OSVehicle) focuses on development of an open source hardware car platform named TABBY EVO [bib#144] for:

> *"electric vehicles prototypes/pre-series/manufacturing … The platform includes all the mechanical and electrical components that make the vehicle move, including chassis, steering and suspension system and complete powertrain … TABBY EVO is shipped in components so to optimize logistics dimensions and import taxes … The platform components and structural design are compliant to road legal certification in the EU, US, and ASIA."*

Local Motors is:

> *"a ground mobility company focused on shaping the future for the better. Founded in 2007 with a belief in open collaboration and co-creation, Local Motors began low volume vehicle manufacturing of open-source designs using multiple micro-factories"*

A recent vehicle developed by Local Motors is Olli [bib#145] which is "the First Co-Created, Self-Driving Vehicle". A picture of one version of Olli [bib#146] is shown below:

picture of one version of Olli [bib#146]

Summary

A vehicle as a component of emerging, disruptive technologies is not new. There is great interest and general public awareness of the paradigm shift in new forms of transportation and new ways to fuel that transportation (e.g., fast trains, electric cars, cars which run on natural gas).

A large, less noticed change is the relationship of vehicles to a major demographic paradigm shift happening in the United States with respect to how people are choosing to live. At least one million Americans have abandoned permanent, traditional (fixed) sheltering (housing) in favor of residence (sheltering) in a vehicle. This vehicle resident *"lifestyle technology"* can be considered emerging and disruptive. It is a paradigm shift. It is a *"new normal"* option for many people.

> *The purpose of this book is to generate discussion of a Sheltering Vehicle Minimal Viable Product (MVP [bib#1]) which supports the "new normal" paradigm shift in housing/sheltering options. The Sheltering Vehicle supports the emerging, disruptive Vehicle Resident lifestyle and is envisioned to be, in and of itself, an emerging, disruptive technology. The intended outcome of this book is the prototyping of a Sheltering Vehicle MVP and the supporting infrastructure.*

A possible new parking option for such a Sheltering Vehicle might be the parking lot of a currently abandoned/vacant mall. Community resources (support facilities, amenities such as bathrooms, showers, washing machines) for those Vehicle Residents in the Sheltering Vehicles could be located in the abandoned mall building or in the parking lot.

Clearly, many people are currently living in vehicles (e.g., RVs, SUVs), however, the lifestyle technology paradigm shift most likely requires the development of an additional vehicle – a Sheltering Vehicle – one that contains the amenities of a shelter, not necessarily the amenities of permanent, traditional housing. Some suggested characteristics of such a Sheltering Vehicle are:

1. Relatively inexpensive to purchase (e.g., $7500 to buy) or rent/lease (can be purchased/leased/rented by individuals or by organizations on behalf of individuals such as organizations and government agencies which address homelessness)

2. Local traffic/road capable mainly (no need for the vehicle to be able to achieve highway speeds for extended periods) and street legal

3. Gas, electric, or hybrid powered, but inexpensive to operate (e.g., high MPG)

4. Easy and inexpensive to maintain (e.g. like the old Volkswagen Beetle, almost any owner could do basic maintenance)

5. Easy to drive (can be an autonomous vehicle)

6. Structured so that at least one person who is at least 6 feet tall can sleep comfortably

7. Some storage capability (e.g., for clothes)

8. Safe and secure for sleeping and stored items

9. No need for incorporated non-shelter amenities (e.g., no cooking facilities, no bathroom)

10. A new option for current and potential Vehicle Residents (is not intended to replace current options)

11. Aesthetically appealing (is not an eyesore in any community in which it is parked)

12. Unique, easily identifiable design characteristics to build community (e.g., similar to the iconic Airstream Community [bib#11]) and deter theft (e.g., similar to the design deterrent of bike-sharing bikes [bib#12])

13. Appealing to both those seeking to be a Vehicle Resident as well as those who are looking for a versatile, functional vehicle with no intention of being a Vehicle Resident (potential market both Vehicle Residents and those living in permanent, traditional housing)

It is hoped that this book provides some inspiration to those interested in the development of a Sheltering Vehicle MVP. It is hoped that this inspiration leads to the completion of a Sheltering Vehicle MVP prototype. But if more inspiration is needed – in terms of the process – the Volkswagen commercial for the new electric car seems to be quite fitting: *Hello Light | ID. BUZZ* [bib#147].

Current Contributors to This Book and Working Group (Feel Free to Distribute, Contribute, and Crowdsource)

As stated, the purpose of this book is to generate discussion of a Sheltering Vehicle Minimal Viable Product (MVP [bib#1]) with the intended outcome of this book being the prototyping of a Sheltering Vehicle MVP and the supporting infrastructure.

As such feel free to crowdsource and distribute this book to anyone you think might be interested in the general topics discussed in this book. Feel free to contribute to this book by adding information and adapting this book as you see fit. As you do this, it is probably worthwhile to keep a list of those who have contributed to the book and those interested in the topic in case anyone in the process wants to follow-up with them in any way. A list is started below:

Current Contributors to this Book and Working Group

- Margaret Kilduff, Ph.D, Professor Emerita, Rutgers University and Associated Sheltering Vehicle Working Group Members. Contact Information: meg.kilduff@rutgers.edu or margaret.kilduff@gmail.com

References Cited in the Book

1. Book Text: *MVP* Link Associated with Text: https://en.wikipedia.org/wiki/Minimum_viable_product
2. Book Text: *Creative Commons Attribution-ShareAlike 4.0 International License* Link Associated with Text: https://creativecommons.org/licenses/by-sa/4.0/
3. Book Text: *following thought* Link Associated with Text: https://en.wikiquote.org/wiki/Jurassic_Park_(novel)
4. Book Text: *The Structure of Scientific Revolutions* Link Associated with Text: https://www.amazon.com/Structure-Scientific-Revolutions-50th-Anniversary-ebook/dp/B007USH7J2/ref=mt_kindle?_encoding=UTF8&me=
5. Book Text: *A man will be imprisoned in a room* Link Associated with Text: https://www.goodreads.com/author/quotes/7672.Ludwig_Wittgenstein
6. Book Text: *"Wittgenstein His Life and Philosophy"* Link Associated with Text: https://www.youtube.com/watch?v=TNaBRR-XeAs#t=23m52s
7. Book Text: *Campground* Link Associated with Text: https://www.bing.com/images/search?view=detailV2&ccid=bjDcbNcl&id=C5C2BB21774725A611DB7F49AD6724A2B2C651AE&thid=OIP.bjDcbNclDi2vxScBf01twAHaER&mediaurl=http%3a%2f%2fflamborovalley.com%2fgallery%2ffun%2fAerial+View+of+Campground.jpg&exph=260&expw=450&q=aerial+view+of+campground&simid=607993250016134540&selectedIndex=0&ajaxhist=0
8. Book Text: *Oregon Trail'R Teardrop Trailer* Link Associated with Text: https://www.bing.com/images/search?view=detailV2&ccid=BH9fDR2p&id=66161A32EF24C981B253ADB2DA15F827834FA4E5&thid=OIP.BH9fDR2p-cVUSAhY4AtDnAHaE7&mediaurl=https%3a%2f%2fwww.oregontrailer.net%2fuploads%2f4%2f6%2f9%2f4%2f46942051%2fot-taylorlake-web-71_orig.jpg&exph=733&expw=1100&q=teardrop+trailers&simid=608012092056207671&selectedIndex=34
9. Book Text: *rats, typhus* Link Associated with Text: https://www.contagionlive.com/news/typhus-outbreak-strikes-los-angeles-but-the-city-is-not-alone-public-health-watch

10. Book Text: *Elio Motors $7,450 Base Price* Link Associated with Text: https://commons.wikimedia.org/wiki/File:Elio-motors-sss-1024x625.jpg
11. Book Text: *Airstream Community* Link Associated with Text: https://www.airstream.com/heritage/why-airstream/people/
12. Book Text: *bike-sharing bikes* Link Associated with Text: https://www.cleveland.com/metro/2014/09/bicycle_share_programs_protect.html
13. Book Text: *one million people* Link Associated with Text: http://rvlife.com/modern-rvers/
14. Book Text: *HUD* Link Associated with Text: https://www.hud.gov/
15. Book Text: *AHAR* Link Associated with Text: https://www.hudexchange.info/homelessness-assistance/ahar/#2018-reports
16. Book Text: *2018 report* Link Associated with Text: https://files.hudexchange.info/resources/documents/2018-AHAR-Part-1.pdf
17. Book Text: *on any given night* Link Associated with Text: https://www.hudexchange.info/news/hud-reports-homelessness-unchanged-in-u-s-in-2018-with-notable-declines-among-veterans-and-families-with-children/
18. Book Text: *$4.75 billion* Link Associated with Text: https://www.everycrsreport.com/reports/RL30442.html#_Toc527712492
19. Book Text: *Homeless Assistance Grants* Link Associated with Text: https://www.hud.gov/program_offices/comm_planning/about/budget
20. Book Text: *$1.8 billion* Link Associated with Text: https://www.va.gov/budget/products.asp
21. Book Text: *homeless veterans* Link Associated with Text: https://www.va.gov/HOMELESS/for_homeless_veterans.asp
22. Book Text: *veterans at-risk* Link Associated with Text: https://www.va.gov/HOMELESS/for_at_risk_veterans.asp
23. Book Text: *Interagency Council on Homelessness* Link Associated with Text: https://www.usich.gov/
24. Book Text: *Homelessness in the United States, Wikipedia* Link Associated with Text: https://en.wikipedia.org/wiki/Homelessness_in_the_United_States
25. Book Text: *National Alliance to End Homelessness* Link Associated with Text: https://endhomelessness.org/

26. Book Text: *Coalition for the Homeless* Link Associated with Text: https://www.coalitionforthehomeless.org/
27. Book Text: *Salvation Army, Provide Shelter* Link Associated with Text: https://www.salvationarmyusa.org/usn/provide-shelter/
28. Book Text: *List of homelessness organizations, Wikipedia* Link Associated with Text: https://en.wikipedia.org/wiki/List_of_homelessness_organizations
29. Book Text: *Is 'Homeless' The Right Word For Those Living On The Street?* Link Associated with Text: https://hoodline.com/2016/12/is-homeless-the-right-word-for-those-living-on-the-street
30. Book Text: *The Homeless Industrial Complex* Link Associated with Text: https://californiaglobe.com/uncategorized/the-homeless-industrial-complex/
31. Book Text: *Disguising a Public Safety Crisis as a Housing Crisis* Link Associated with Text: https://humanevents.com/2019/06/12/disguising-a-public-safety-crisis-as-a-housing-crisis/
32. Book Text: *choose this option* Link Associated with Text: https://slate.com/business/2018/08/vehicular-homelessness-is-on-the-rise-should-cities-help-people-sleep-in-their-cars.html
33. Book Text: *students are homeless* Link Associated with Text: https://sanjosespotlight.com/more-students-face-homelessness-as-sjsu-school-year-ends/
34. Book Text: *highest number of homeless students* Link Associated with Text: https://sanjosespotlight.com/sjsu-students-rally-for-homeless-students-rights/
35. Book Text: *Single Room Occupancy (SRO)* Link Associated with Text: https://en.wikipedia.org/wiki/Single_room_occupancy
36. Book Text: *Starcity* Link Associated with Text: https://www.curbed.com/2018/3/8/17097154/affordable-housing-coliving-starcity-common-sro
37. Book Text: *Common* Link Associated with Text: https://www.curbed.com/2017/8/11/16135720/coliving-roommate-apartment-millennial
38. Book Text: *WeLive* Link Associated with Text: https://www.curbed.com/2016/10/14/13274082/wework-welive-vice-news-tonight-hbo

39. Book Text: *capsule or pod hotel* Link Associated with Text: https://en.wikipedia.org/wiki/Capsule_hotel
40. Book Text: *First Cabin Tsukiji* Link Associated with Text: https://www.cnn.com/travel/article/tokyo-luxury-capsule-hotels/index.html
41. Book Text: *internet café* Link Associated with Text: https://soranews24.com/2017/06/05/japanese-internet-cafe-chain-evolves-into-internet-hotel-great-for-backpackers-budget-travelers/
42. Book Text: *Net Room* Link Associated with Text: https://www.straitstimes.com/lifestyle/travel/japanese-internet-cafe-chain-transforms-into-budget-hotel-for-travellers
43. Book Text: *backyard sheds* Link Associated with Text: https://sanfrancisco.cbslocal.com/2019/04/04/oakland-to-invest-millions-in-tuff-shed-program-to-help-house-homeless/
44. Book Text: *pop-up shelters* Link Associated with Text: https://www.fastcompany.com/90365347/pallet-shelters-let-cities-offer-quick-housing-to-the-homeless
45. Book Text: *Homes on Wheels: Meet the People Who Live in RVs on San Francisco Streets* Link Associated with Text: https://www.nbcbayarea.com/news/local/Homes-on-Wheels-Meet-the-People-Who-Live-in-RVs-on-San-Francisco-Streets-503418641.html
46. Book Text: *More Americans are forced to "reside" in their vehicles* Link Associated with Text: https://www.cbsnews.com/news/more-americans-are-living-in-their-vehicles-amid-high-housing-prices/
47. Book Text: *America's Housing Crisis Is Forcing More People To Live In Vehicles* Link Associated with Text: https://www.huffpost.com/entry/america-housing-crisis-living-in-vehicles_n_5b680be3e4b0fd5c73db49b
48. Book Text: *'It's the New Form of Affordable Housing': More People Are Living in Their Cars* Link Associated with Text: https://www.governing.com/topics/health-human-services/gov-homeless-shelter-car.html
49. Book Text: *For Countless Americans, Living In Their Vehicles Has Become "The New Normal"* Link Associated with Text: https://www.zerohedge.com/news/2019-06-12/countless-americans-living-their-vehicles-has-become-new-normal

50. Book Text: *Mountain View Vehicle Residents throw party for public* Link Associated with Text: https://mv-voice.com/news/2019/06/24/mountain-view-vehicle-residents-throw-party-for-public
51. Book Text: *San Jose State University has highest number of homeless students across CSU system* Link Associated with Text: https://abc7news.com/education/sjsu-has-highest-number-of-homeless-students-across-csu-system/4849381/
52. Book Text: *Homeless Shelter* Link Associated with Text: https://en.wikipedia.org/wiki/Homeless_shelter
53. Book Text: *Tent City, Wikipedia* Link Associated with Text: https://en.wikipedia.org/wiki/Tent_city
54. Book Text: *Is It a Good Idea to Give Tents to Homeless People?* Link Associated with Text: https://ecobear.co/homeless-encampment-cleaning/give-tents-homeless/
55. Book Text: *Tents-4-Homeless* Link Associated with Text: http://www.tents4homeless.org/
56. Book Text: *Weedpatch Camp* Link Associated with Text: https://en.wikipedia.org/wiki/Weedpatch_Camp
57. Book Text: *it is estimated* Link Associated with Text: https://wandrlymagazine.com/article/rv-camping-cost-per-night/
58. Book Text: *Go RVing* Link Associated with Text: https://gorving.com/where-to-go
59. Book Text: *RV Parks and Campgrounds* Link Associated with Text: https://gorving.com/where-to-go/rv-parks--campgrounds
60. Book Text: *public land campgrounds* Link Associated with Text: https://gorving.com/where-to-go/public-lands
61. Book Text: *cater to RVers* Link Associated with Text: https://en.wikipedia.org/wiki/RV_park
62. Book Text: *New Beginnings Counseling Center* Link Associated with Text: https://sbnbcc.org/safe-parking/
63. Book Text: *Dreams for Change* Link Associated with Text: https://www.dreamsforchange.org/the-safe-parking-program/
64. Book Text: *Safe Parking LA* Link Associated with Text: https://www.safeparkingla.org/about-spala

65. Book Text: *AB-302 Parking: homeless students* Link Associated with Text: https://leginfo.legislature.ca.gov/faces/billTextClient.xhtml?bill_id=201920200AB302
66. Book Text: *2 billion parking spaces* Link Associated with Text: https://www.citylab.com/transportation/2018/11/parking-lots-near-me-shopping-plazas-vacant-spaces/576646/
67. Book Text: *dead mall* Link Associated with Text: https://en.wikipedia.org/wiki/Dead_mall
68. Book Text: *450 dead malls* Link Associated with Text: https://en.wikipedia.org/wiki/Deadmalls.com
69. Book Text: *California bill would let homeless college students sleep in cars on campus amid housing crisis* Link Associated with Text: https://www.foxnews.com/politics/california-bill-would-let-homeless-college-students-sleep-in-cars-on-campus-amid-housing-crisis
70. Book Text: *50 Unconventional Free Overnight Parking Recommendations* Link Associated with Text: https://www.truckcampermagazine.com/question-of-the-week/50-unconventional-free-overnight-parking-recommendations/
71. Book Text: *8 Places to Park and Sleep in Your Car Overnight for Free (Legally)* Link Associated with Text: https://firstquarterfinance.com/places-to-park-and-sleep-in-you-car-overnight-for-free/
72. Book Text: *All About Overnight Parking* Link Associated with Text: https://divineontheroad.com/overnight-parking/
73. Book Text: *Vanlifer's Guide to Sleeping, Camping, and Overnight Parking* Link Associated with Text: https://gnomadhome.com/vanlife-sleeping-camping-overnight-parking/
74. Book Text: *Finding Home in a Parking Lot* Link Associated with Text: https://www.citylab.com/equity/2019/02/homeless-safe-parking-lots-sleeping-in-cars-city-programs/581128/
75. Book Text: *Living Behind the Wheel* Link Associated with Text: https://slate.com/business/2018/08/vehicular-homelessness-is-on-the-rise-should-cities-help-people-sleep-in-their-cars.html
76. Book Text: *Where to Sleep in Your Car When Homeless* Link Associated with Text: http://homelessadvice.com/where-to-sleep-in-your-car-when-homeless/

77. Book Text: *Parking Has Eaten American Cities* Link Associated with Text: https://www.citylab.com/transportation/2018/07/parking-has-eaten-american-cities/565715/
78. Book Text: *Quantified Parking: Comprehensive Parking Inventories for Five U.S. Cities* Link Associated with Text: https://www.mba.org/Documents/Research/RIHA/18806_Research_RIHA_Parking_Report.pdf
79. Book Text: *List of shopping malls in the United States* Link Associated with Text: https://en.wikipedia.org/wiki/List_of_shopping_malls_in_the_United_States
80. Book Text: *Hidden in Plain Sight: Finding Safe Parking for Vehicle Residents* Link Associated with Text: https://papers.ssrn.com/sol3/papers.cfm?abstract_id=3173221
81. Book Text: *Championing Seattle's Invisible Homeless* Link Associated with Text: https://artsci.washington.edu/news/2012-12/championing-seattles-invisible-homeless
82. Book Text: *Seattle's forgotten street community: UW anthropologist talks about the unique circumstances of vehicle residency* Link Associated with Text: http://www.washington.edu/news/2019/05/31/seattles-forgotten-street-community-uw-anthropologist-talks-about-unique-circumstances-of-vehicle-residency/
83. Book Text: *Seattle Vehicular Residency Research Project 2012 Advisory Report* Link Associated with Text: https://www.google.com/url?sa=t&rct=j&q=&esrc=s&source=web&cd=3&cad=rja&uact=8&ved=2ahUKEwid3YHs-IfjAhWGLs0KHZ1-DBQQFjACegQIARAC&url=http%3A%2F%2Fclerk.seattle.gov%2F~public%2Fmeetingrecords%2F2012%2Fhhshc20120926_8a.pdf&usg=AOvVaw238aoc_sZWNVUyFIeqYE8e
84. Book Text: *EscapeesRVClub* Link Associated with Text: https://www.escapees.com/
85. Book Text: *establish a domicile* Link Associated with Text: https://www.escapees.com/education/domicile/
86. Book Text: *mail forwarding service* Link Associated with Text: https://www.escapees.com/mail-service/how-escapees-mail-forwarding-service-works/

87. Book Text: *internet options* Link Associated with Text: https://www.escapees.com/comparing-our-rv-internet-options/
88. Book Text: *have a cell phone* Link Associated with Text: https://www.calhealthreport.org/2019/01/11/expired-lost-stolen-cell-phones-critical-homeless-people-can-tough-get-keep/
89. Book Text: *HomelessnessAdvice.com: Advice When You Need It, Guides* Link Associated with Text: http://homelessadvice.com/category/guides/
90. Book Text: *Vanlife How To: Your Complete Guide to Life on the Road* Link Associated with Text: https://gnomadhome.com/vanlife-how-to/
91. Book Text: *Workamper News* Link Associated with Text: https://workamper.com/
92. Book Text: *How to Establish State Residency if you Live in Your RV Full Time* Link Associated with Text: https://rvshare.com/blog/state-residency/
93. Book Text: *Domicile for Nomads: Establishing a Legal Residence on the Road* Link Associated with Text: https://gnomadhome.com/domicile-for-nomads/
94. Book Text: *Mail and Packages on the Road: A Guide for Vanlife Nomads* Link Associated with Text: https://gnomadhome.com/vanlife-mail-and-packages-for-nomads/
95. *Helping you Live a Mobile & Connected Lifestyle!*
96. Book Text: *RV Internet Access* Link Associated with Text: https://www.your-rv-lifestyle.com/rv-internet-access/
97. Book Text: *guides for Nomads/Travelers* Link Associated with Text: https://thevanual.com/health-and-hygiene/
98. Book Text: *gym membership* Link Associated with Text: https://divineontheroad.com/van-life-best-gym-memberships/
99. Book Text: *Planet Fitness* Link Associated with Text: https://www.planetfitness.com/
100. Book Text: *24 Hour Fitness* Link Associated with Text: https://www.24hourfitness.com/
101. Book Text: *discounted membership* Link Associated with Text: https://www.costco.com/gym-memberships-wellness.html?brand=24-hour-fitness&refine=ads_fbrand_ntk_cs%253A%252224%2BHour%2BFitness%2522%7C
102. Book Text: *Marin Mobile Care: The Showers* Link Associated with Text: Marin Mobile Care: The Showers

103. Book Text: *Lava Mae Mobile Hygiene* Link Associated with Text: https://lavamae.org/
104. Book Text: *Dignity on Wheels (Showers and Laundry)* Link Associated with Text: https://www.dignityonwheels.org/
105. Book Text: *The Laundry Truck LA* Link Associated with Text: https://www.thelaundrytruckla.com/
106. Book Text: *Homeward Alliance Mobile Laundry* Link Associated with Text: https://www.homewardalliance.org/mobile-laundry/
107. Book Text: *Revive and Refresh* Link Associated with Text: https://www.revive-refresh.org/
108. Book Text: *Homeless Encampment Cleaning Company* Link Associated with Text: https://ecobear.co/homeless-encampment-cleaning/
109. Book Text: *Portable Restroom Trailers Sales and Rentals* Link Associated with Text: https://www.portablerestroomtrailers.com/
110. Book Text: *Mobile Bathroom* Link Associated with Text: https://www.scanvogn.com/products/mobile-bathroom
111. Book Text: *Luxury Restroom Trailers* Link Associated with Text: http://onsiteco.com/luxury-restroom-trailers/
112. Book Text: *PortaPotty Rental Guide* Link Associated with Text: http://www.portapotty.net/rental/
113. Book Text: *Dumpsters.com* Link Associated with Text: https://www.dumpsters.com/sizes/commercial-dumpster-sizes
114. Book Text: *for a self-storage unit* Link Associated with Text: https://www.sparefoot.com/self-storage/news/1432-self-storage-industry-statistics/
115. Book Text: *soup kitchen* Link Associated with Text: https://en.wikipedia.org/wiki/Soup_kitchen
116. Book Text: *food bank* Link Associated with Text: https://en.wikipedia.org/wiki/Food_bank
117. Book Text: *Temporary Kitchens 123* Link Associated with Text: https://temporarykitchens123.com/
118. Book Text: *U.S. Mobile Kitchens* Link Associated with Text: https://www.usmobilekitchens.com/
119. Book Text: *10 Best Cargo and Carriers for Cars in 2019* Link Associated with Text: https://mycarneedsthis.com/todays-best-cargo-boxes-and-top-carriers/

120. Book Text: *Thule Rooftop Cargo Carriers* Link Associated with Text: https://www.thule.com/en-us/us/cargo-carrier/car-top-carrier
121. Book Text: *Home Depot Storage Bins* Link Associated with Text: https://www.homedepot.com/b/Storage-Organization-Storage-Containers-Storage-Bins/N-5yc1vZcl3z
122. Book Text: *many RV options* Link Associated with Text: https://gorving.com/compare-rvs
123. Book Text: *campervan* Link Associated with Text: https://en.wikipedia.org/wiki/Campervan
124. Book Text: *Volkswagen Autosleeper Clubman GL* Link Associated with Text: https://en.wikipedia.org/wiki/File:Clubman.jpg
125. Book Text: *tiny houses* Link Associated with Text: https://www.tinyhomebuilders.com/tiny-houses
126. Book Text: *teardrop trailer* Link Associated with Text: https://en.wikipedia.org/wiki/Teardrop_trailer
127. Book Text: *England's Teardrop Trailers* Link Associated with Text: http://www.teardroptrailers.co.uk/
128. Book Text: *nuCamp* Link Associated with Text: https://nucamprv.com/
129. Book Text: *example of a teardrop trailer* Link Associated with Text: https://commons.wikimedia.org/wiki/File:Turtle1500TeardropTrailer.jpg
130. Book Text: *Ford Transit Connect Cargo Van* Link Associated with Text: https://www.ford.com/commercial-trucks/transit-connect-cargo-van/?gnav=header-all-vehicles
131. Book Text: *Ford Transit Connect Passenger Wagon* Link Associated with Text: https://www.ford.com/trucks/transit-connect-passenger-van-wagon/?gnav=header-all-vehicles
132. Book Text: *Van Life Guide: How To Build A DIY Camper Van Conversion* Link Associated with Text: https://www.parkedinparadise.com/how-to-live-in-van/
133. Book Text: *The Vanual Build Manual* Link Associated with Text: https://thevanual.com/
134. Book Text: *Build Your Van* Link Associated with Text: https://gnomadhome.com/build-your-van/
135. Book Text: *DIY camper van: 5 affordable conversion kits you can buy now* Link Associated with Text:

https://www.curbed.com/2018/8/1/17640472/camper-van-vanlife-diy-kit-for-sale

136. Book Text: *Campervan Conversion Kit: 7 simple ways to DIY your van* Link Associated with Text: https://www.thewaywardhome.com/campervan-conversion-kit/
137. Book Text: *Ursa Minor Vehicles* Link Associated with Text: http://www.ursaminorvehicles.com/
138. Book Text: *research vehicles from Volkswagen* Link Associated with Text: https://www.volkswagenag.com/en/group/research/research-vehicles.html
139. Book Text: *Volkswagen NILS* Link Associated with Text: https://commons.wikimedia.org/wiki/File:Volkswagen_Nils_(side).jpg
140. Book Text: *Elio Motors* Link Associated with Text: https://www.eliomotors.com/
141. Book Text: *Elio Motors three-wheel vehicle* Link Associated with Text: https://commons.wikimedia.org/wiki/File:ElioSideView1.jpg
142. Book Text: *Local Motors* Link Associated with Text: https://localmotors.com/
143. Book Text: *Open Motors* Link Associated with Text: https://www.openmotors.co/homepage/
144. Book Text: *TABBY EVO* Link Associated with Text: https://www.openmotors.co/product/tabby-evo/
145. Book Text: *Olli* Link Associated with Text: https://localmotors.com/meet-olli/
146. Book Text: *picture of one version of Olli* Link Associated with Text: https://en.wikipedia.org/wiki/File:Goodyear_Olli_Genf_2019_1Y7A5023.jpg
147. Book Text: *Hello Light | ID. BUZZ* Link Associated with Text: https://www.youtube.com/watch?v=qEvNL6oEr0U

Made in the USA
Monee, IL
05 November 2026

67375663R00033